# Animals

© 2015 OnBoard Academics, Inc
Portsmouth, NH
800-596-3175
www.onboardacademics.com
ISBN: 978-1-63096-061-2

OnBoard Academic's books are specifically designed to be used as printed workbooks or as on-screen instruction. Each page offers focused exercises and students quickly master topics with enough proficiency to move on to the next level.

OnBoard Academic's lessons are used in over 25,000 classrooms to rave reviews. Our lessons are aligned to the most recent governmental standards and are updated from time to time as standards change. Correlation documents are located on our website. Our lessons are created, edited and evaluated by educators to ensure top quality and real life success.

Interactive lessons for digital whiteboards, mobile devices, and PCs are available at www.onboardacademics.com. These interactive lessons make great additions to our books.

You can always reach us at customerservice@onboardacademics.com.

# Animal Life Cycle

Match each animal with its infant.

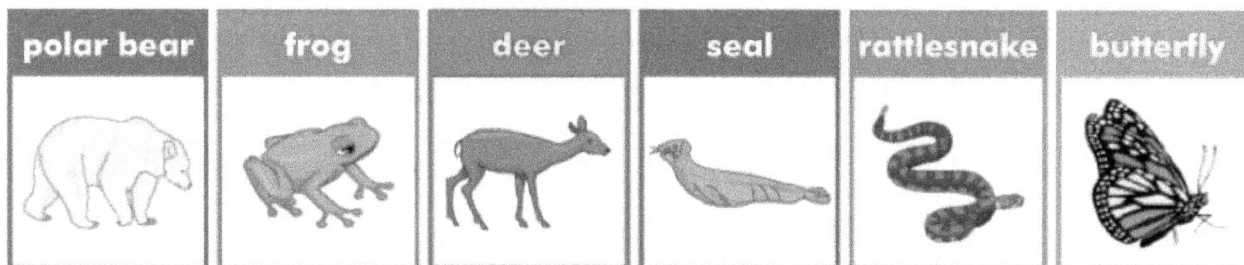

| polar bear | frog | deer | seal | rattlesnake | butterfly |
|------------|------|------|------|-------------|-----------|

What is a life cycle.

A life cycle is the name we give to the set of stages that a living creature goes through.

When human beings are born, we start out as babies, then we grow and develop into toddlers and then a bit later into kids. At the age of 13 we are officially teenagers and then become adults. We are adults for the longest period in our life cycle. After adulthood we become seniors.

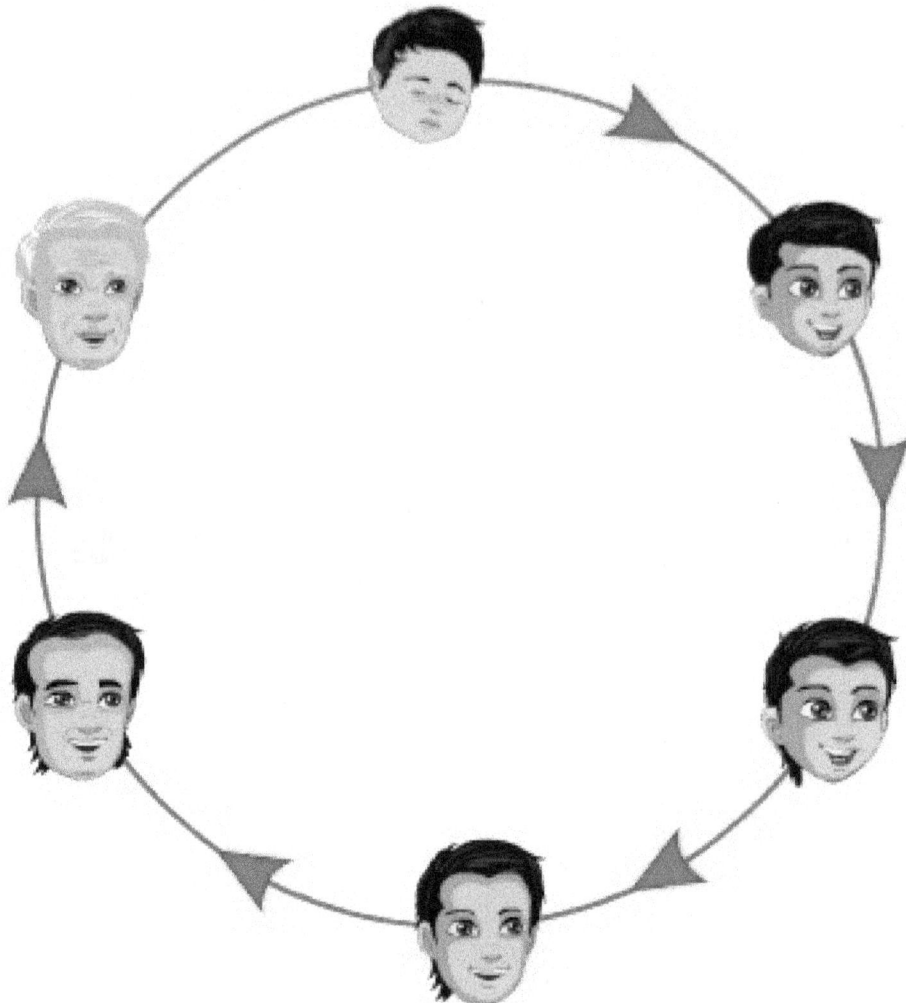

Life cycles of a chicken, a dog and a dragonfly.

All birds including chickens start their life as an embryo within an egg. The embryo take about 21 days to grow into a chick and hatch. The yolk of an egg is the food source and the hard outer shell of the egg protects the embryo. The chick uses its beak to break out of the shell and sometimes it takes a whole day. The chicks first feathers are down and it takes about 4 weeks to grow its outdoor feathers. It takes about 6 months for the chick to grow into an adult chicken. Female chickens are called hens and male chickens are called roosters.

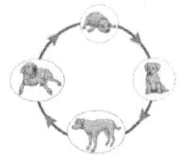

| Egg | Embryo | Chick | Hen | Rooster |
|-----|--------|-------|-----|---------|

A puppy is born in a litter often with about half a dozen other puppies. Newborn puppies are born with their eyes close and don't normally open their eyes for 2 weeks. During the first few months the puppy will grow quickly and become bigger and heavier. After six weeks it will no longer need its mother's milk. At about two years the puppy will be a fully grown dog and continue the life cycle by having its own puppies although dogs can have puppies at just one year old. As dogs age, the hair around their mouth often turns white or gray. Dogs usually live about 12 years but some will live longer.

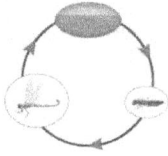

All dragonflies start out as eggs. The mother dragonfly lays the eggs in the water. Some of the eggs will hatch in about 5 days but others will take months to hatch. At this stage the dragonfly is a larva or nymph which is the second stage of a dragonfly's life cycle. This is the longest stage of a dragonfly's life and spent entirely under water lasting up to four years. Eventually the larva grows wings and turns into a dragonfly and no longer lives under water. The adult stage of a dragonfly's life is quite brief lasting only a couple of months. During this time the dragonfly spends its time looking for a mate to continue the life cycle.

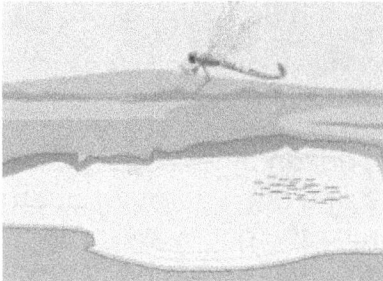

Complete this dog's lifecycle with the suggestions and illustrations below.

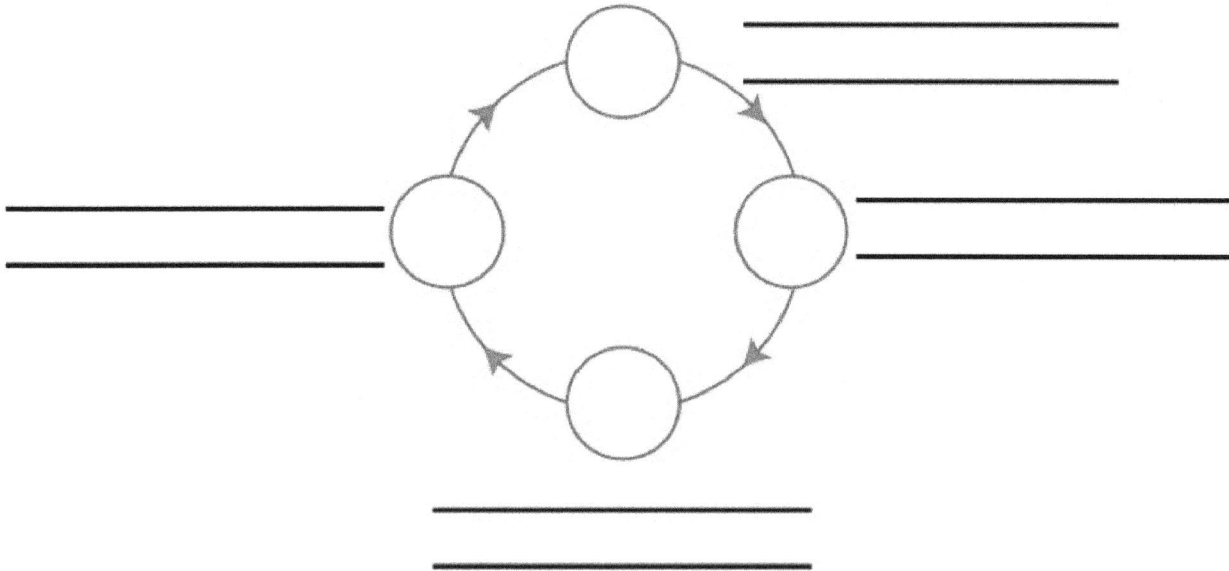

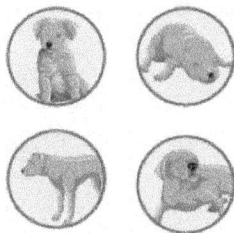

adult

hair turns white

stops drinking milk

eyes are closed

fully grown

old adult

puppy

newborn

Complete the dragonfly's life cycle.

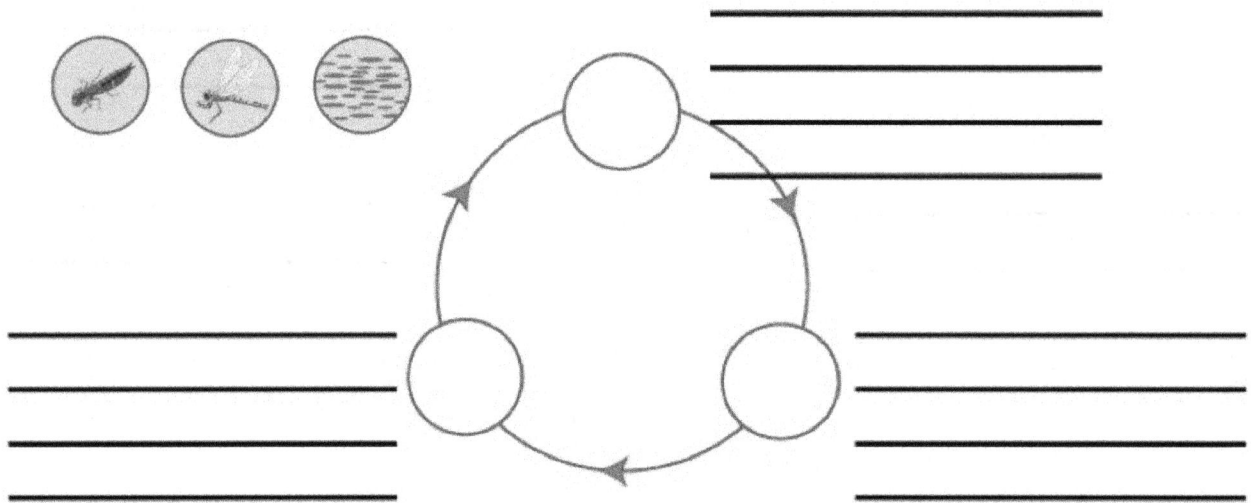

_____
_____
_____
_____

_____          _____
_____          _____
_____          _____
_____          _____

about four years          lives in air and land          also called nymph

larva          adult          up to a few months

lives in water          lives in still water          may hatch in 5 days

about two months          has wings

egg

Do you know your infant animal names?

| deer | pig | kangaroo | goat |
|---|---|---|---|
| | | | |

| gorilla | goose | whale | tiger |
|---|---|---|---|
| | | | |

shoat            joey            kid            gosling

cub            infant            calf            fawn

What is my average life span?

Can you guess at the average life span of these animals.  Insert the first two letters of the animals name into the circle that you believe represents its life span.

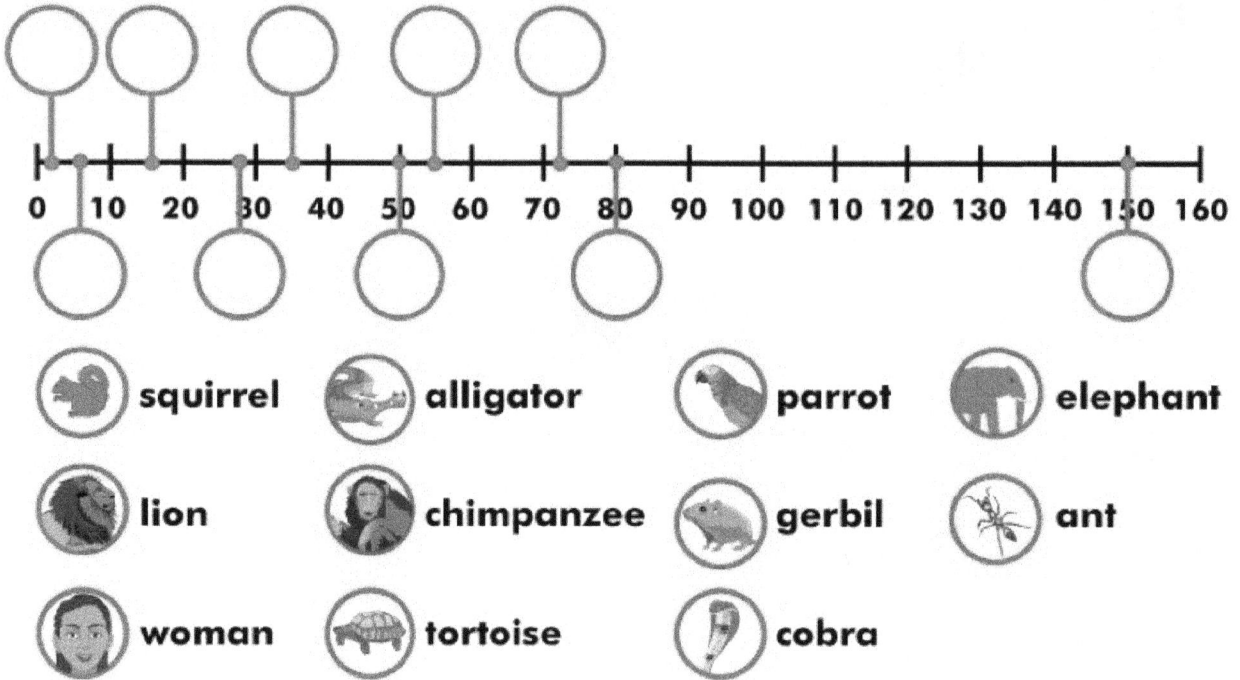

squirrel

alligator

parrot

elephant

lion

chimpanzee

gerbil

ant

woman

tortoise

cobra

All animals eventually die; this is the natural end of a life cycle. What factors do you think influence how long an animal lives on average?

answer

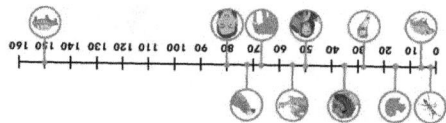

## Animal Life Cycle Quiz

1. The set of stages that a living creature goes through is
   called a _____.
   - a. child cycle
   - b. life cycle
   - c. human cycle

2. How many days does an embryo take to grow into a chick?
   - a. 12
   - b. 22
   - c. 21

3. Female chickens are called roosters.  True or false?

4. A puppy turns into a full grown dog in about _____.
   - a. 1 month
   - b. 6 months
   - c. 24 months

5. The second stage of a dragonfly's life cycle is called

   _____.

6. A dragonfly spends most of its life as a larva.  True or false?

www.onboardacademics.com

# Physical Characteristics of Animals

www.onboardacademics.com

Can you match the animal with its outer covering?

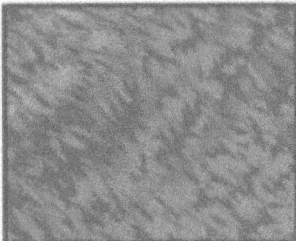

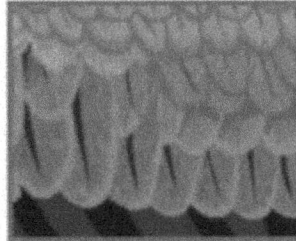

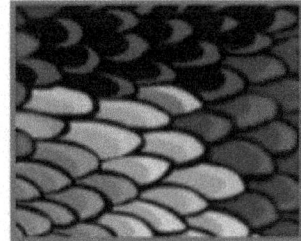

These animals have very different physical characteristics. Physical characteristics describe what an animal looks like, including things like its shape, size, and color, and the type of body parts that it has. For example, birds have feathers, snakes have scales, deer have fur, and snails have hard shells.

www.onboardacademics.com

Different Birds, Different Feathers.

See if you can match the picture and description of the feather type with the correct bird.

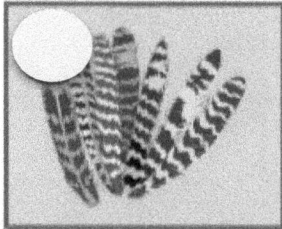

My feathers help me to stay camouflaged so I can hide.

My feathers are designed to be quiet when I fly. This helps me to hunt better.

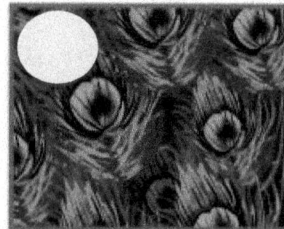

I use my feathers to attract females of my species.

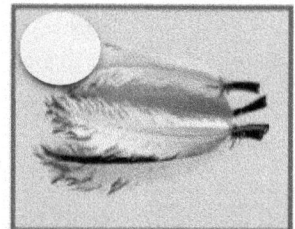

My feathers are short, broad and packed closely together to keep out water and keep in warmth.

**Peacock**

**Penguin**

**Turkey**

**Owl**

www.onboardacademics.com

Different Animals, Different Teeth

See if you can match the picture and the description of the teeth with the proper animal.

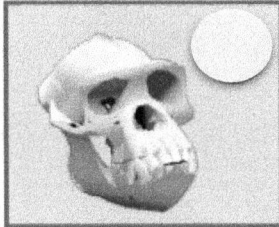
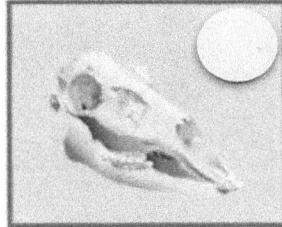
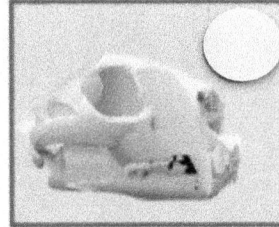
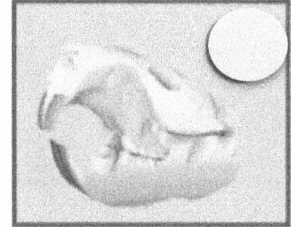

| | | | |
|---|---|---|---|
| I'm an omnivore. I use my incisors to bite into meat and my molars to grind up plants. | My molars are great for eating grass and other plants. I'm great at grinding plants. | My pointy canine teeth are great for hunting animals and tearing meat. | My huge incisor teeth are big and strong. I can even use them to cut trees. |

**Cougar**          **Chimpanzee**          **Beaver**          **Deer**

Different Animals, Different Feet

See if you can match the picture and the description of feet with the correct animal.

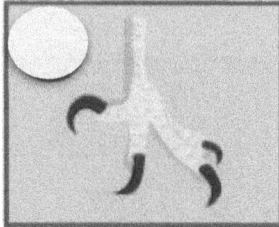

My feet have sharp talons. I use my feet to pick up animals.

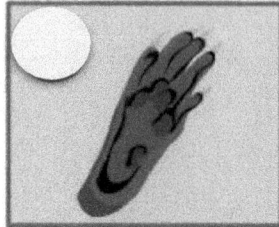

My feet are great at grabbing small branches. I also have long claws to help me climb.

My feet are webbed and help me to swim and maneuver quickly in the water.

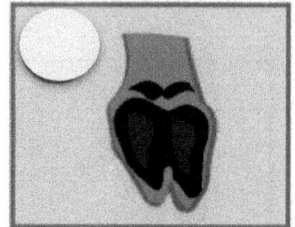

My feet are designed perfectly for climbing and holding onto small ledges.

**Squirrel**

**Platypus**

**Hawk**

**Mountain Goat**

Sort the animals by the number of legs that they have.

| | | |
|---|---|---|
| **0 legs** | **2 legs** | **4 legs** |

Starfish   Camel   Chicken   Whale   Grasshopper

Human   Snake   Butterfly   Goat   Hummingbird

Alligator   Spider   Octopus   Ant   Hedgehog

| | | |
|---|---|---|
| **5 legs** | **6 legs** | **8 legs** |

www.onboardacademics.com

Warm Blooded and Cold Blooded Animals.

Look at the following three illustrations. What happens to the dog and the lizard when the temperature (shown on the left) changes.

Dog: _____

Lizard: _____

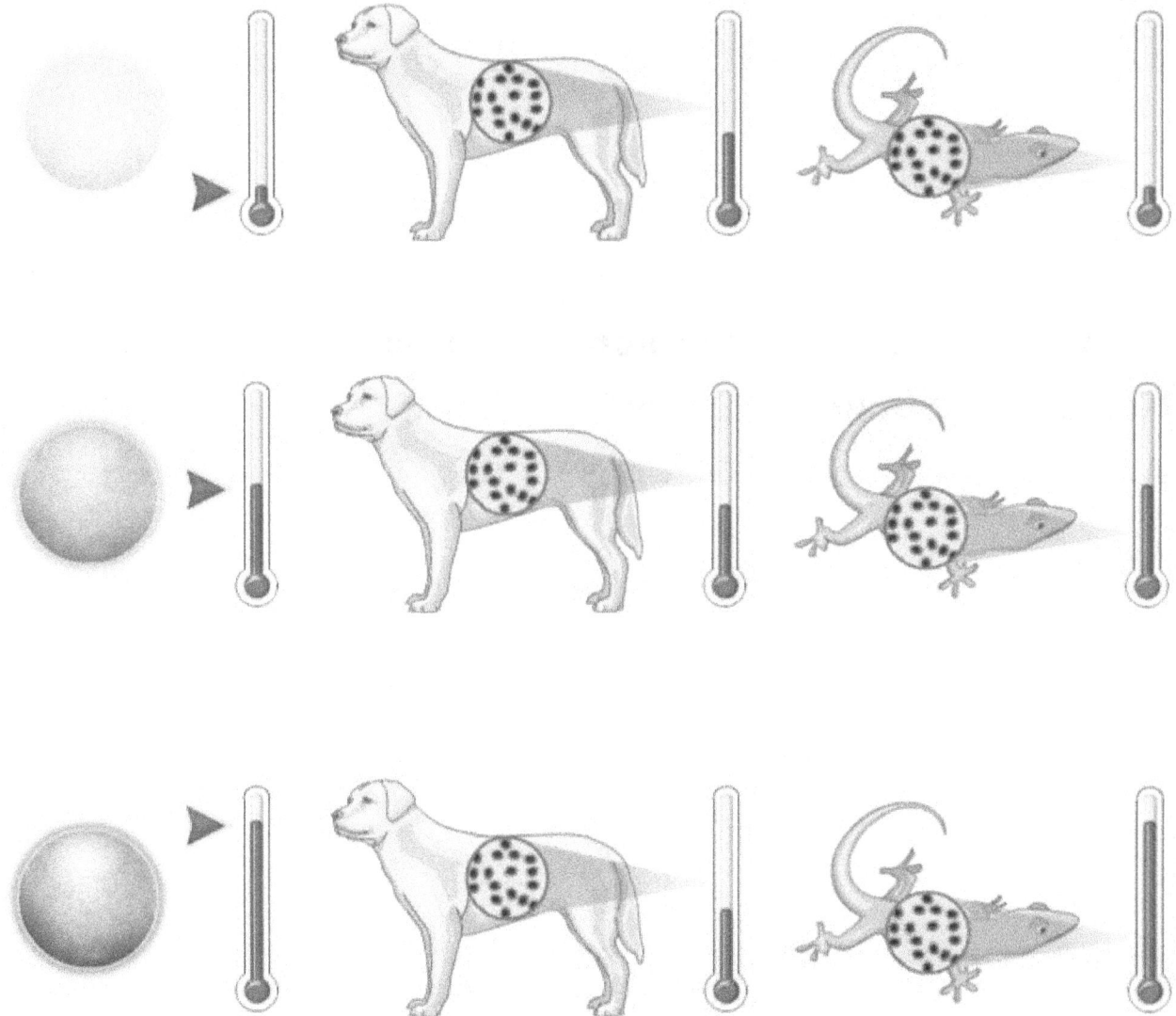

**The blood of warm-blooded animals stays at the same temperature, even when the outside temperature rises or falls. The blood of cold-blooded animals rises and falls with the outside temperature. Cold-blooded animals sometimes have cold blood, and sometimes have warm blood, but warm-blooded animals always have warm blood.**

How do I breath?

All animals use either lungs, gills or spiracles to breath.

All animals need oxygen to breathe. Some animals use gills to get oxygen from the water. Some animals use lungs to get oxygen from air. Other animals have gills that turn into lungs. The smallest animals have spiracles: tiny holes on their bodies that let in oxygen.

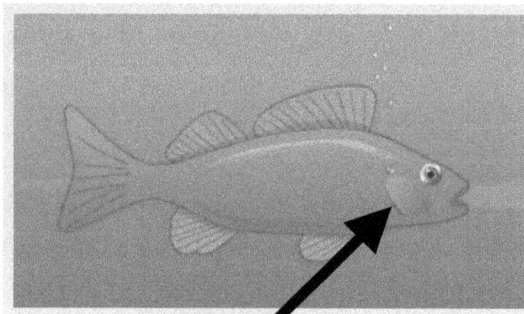

Gills

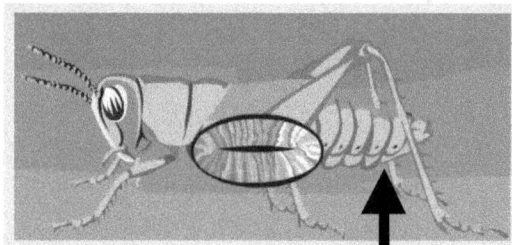

Spiracles

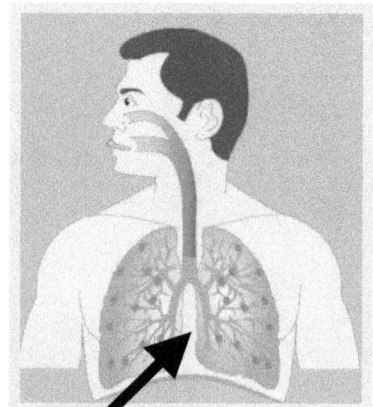

Lungs

www.onboardacademics.com

How do I breath?
Label each animal L for Lungs,  G for Gills or S for Spiracles

| Mammals | Fish | Reptiles |
|---|---|---|
| | | |
| | | |

| Amphibians | Birds | Insects |
|---|---|---|
| | | |
| | | |

## Physical Characteristics of Animals Quiz

1. All animals share the same physical characteristics. True or false?

2. The feathers of _____ help them camouflage from predators.
   - a. turkeys
   - b. peacocks
   - c. eagles

3. Which animal has wide molars to grind plants?
   - a. cougar
   - b. shark
   - c. chimpanzee

4. Beavers have canine teeth for tearing flesh. True of false.

5. A _____ has strong sharp claws for digging tree bark while climbing.
   - a. platypus
   - b. hawk
   - c. squirrel

# Animal Adaptations; Teeth and Diet

     www.onboardacademics.com

Can you guess how many teeth humans have?  Adults have more teeth than children.

**16      20      32      40      64**

Can you guess how many teeth these animals have?

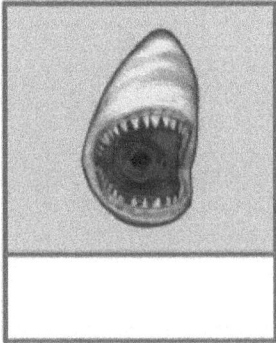

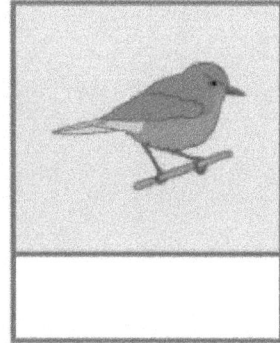

30          300          42          0

**Different types of animals have different types and different numbers of teeth. Some animals, such as insects, birds, turtles, and some fish, don't have any teeth. What an animal eats (its diet) is an important factor in the number and type of teeth that it has.**

How did you do?

 32

 20

 300

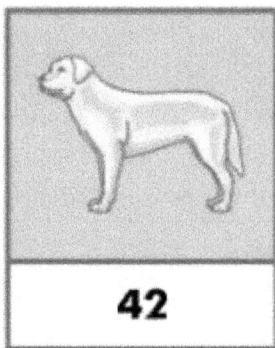 42

 30

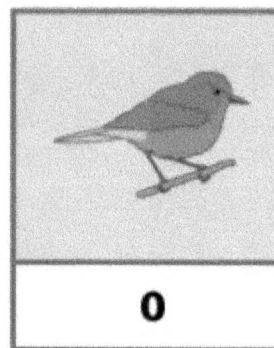 0

www.onboardacademics.com

## Types of Teeth

There are three types of teeth. Incisors are good for cutting and biting, canines are good for tearing, and molars are good for grinding.

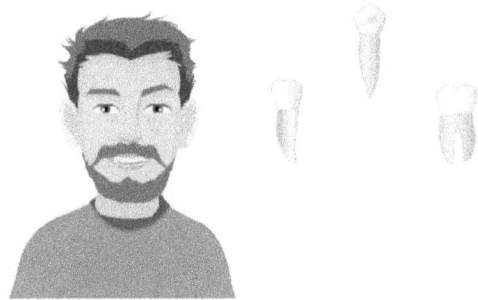

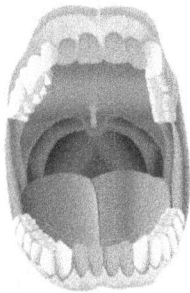

In a human mouth we have 8 incisors, two in the upper jaw and two in the lower jaw. These teeth have a flat thin edge that helps to bite and cut food.

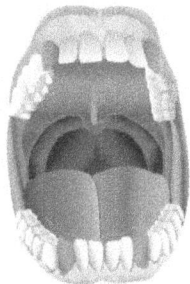

Next to the incisors are the canine teeth. These are long pointed teeth that help us to cut meat. There is a canine tooth on each side of the incisors on the top and bottom for a total of four canine teeth.

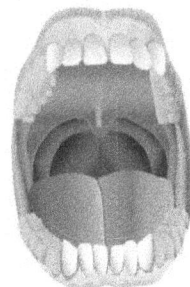

Behind the canine teeth are the molars. Molars are designed to help us to chew and grind food. Children have eight molars and adults have twenty.

Just like humans the type and number of teeth that an animal has is related to its diet. A tiger has very large canine teeth in its upper jaw to bite other animals.

A rabbit doesn't have any canine teeth.

Horses and cows have incisors to help them to cut grass and other plants. They also have strong molars to help them to chew and grind rough vegetation.

Some animals like birds, fish and insects don't have any teeth at all.

Why do you think some animals do not have any teeth? _____

_____

_____

Identify each tooth and match it with its main job.

# canines

# molars

# incisors

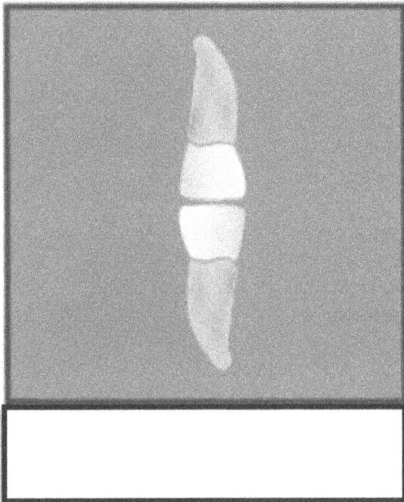

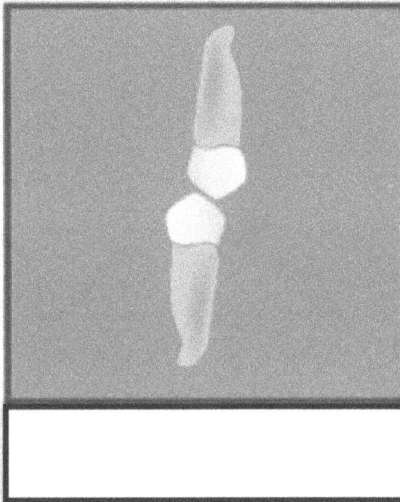

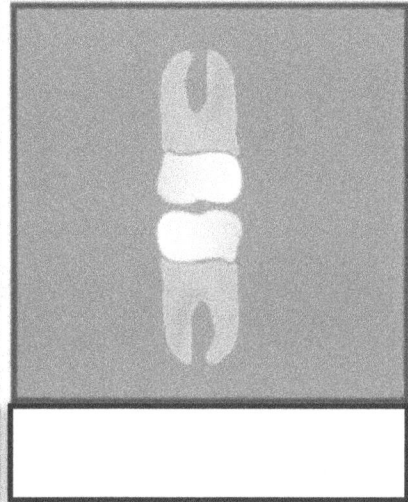

We have four of these teeth.  We use them to tear things like meat.

We have eight of these teeth. They are great for cutting and biting.

We have 20 of these when we are older but only 8 when we are younger.  We use these go grind and chew things.

www.onboardacademics.com

Match these animals with their food.

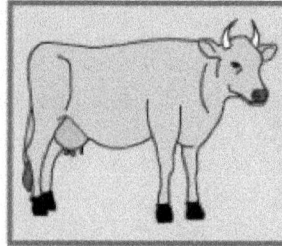

Match these animals with their teeth.

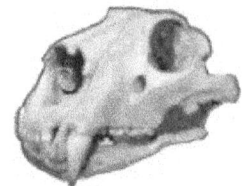

What's for dinner?

Study the skeletons of teeth and look for the different types of teeth that exist in each skeleton.

Place a P for plant and or an M for meat in the boxes below each animal to indicate the type of food it eats based on what you know about its teeth.

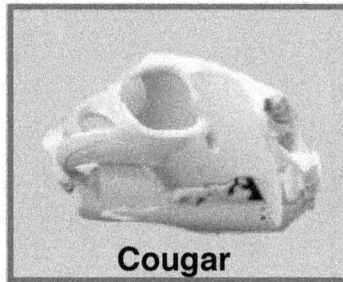

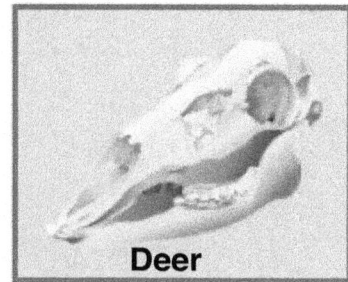

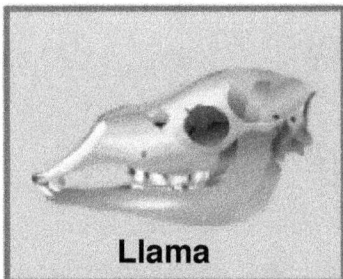

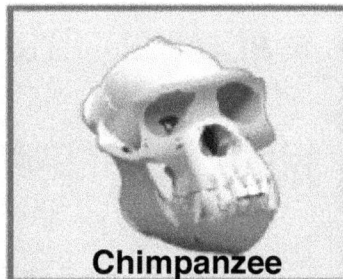

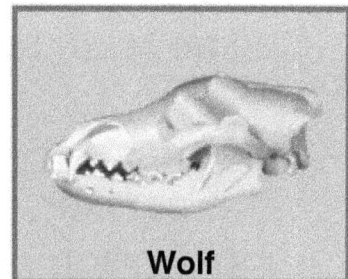

| Shark | Cougar | Deer |
|---|---|---|
| | | |

| Llama | Chimpanzee | Wolf |
|---|---|---|
| | | |

## Animal Adaptation: Teeth and Diet Quiz

1. Insects have teeth.  True or false?

2. All animals have the same number of teeth.  True or false?

3. Teeth that help us to cut and bite food are called _____.
   a. incisors
   b. canines
   c. molars

4. Children have _____ molars.
   a. 5
   b. 20
   c. 8

5. Canine teeth help us tear meat.  True or false?

6. Cows have _____ teeth to cut grass and other plants.
   a. canine
   b. incisor
   c. molar

7. Meat eating animals have strong molars to cut meat.  True or false?

8. Rabbits have sharp canine teeth.  True or false?

www.ingramcontent.com/pod-product-compliance
Lightning Source LLC
Chambersburg PA
CBHW062030210326
41519CB00060B/7386